D0886567

The Jessie and John Danz Lectures

The Jessie and John Danz Lectures

The Human Crisis, by Julian Huxley
Of Men and Galaxies, by Fred Hoyle
The Challenge of Science, by George Boas
Of Molecules and Men, by Francis Crick
Nothing But or Something More, by Jacquetta Hawkes
How Musical Is Man?, by John Blacking
Abortion in a Crowded World:
The Problem of Abortion with Special Reference to India,
by S. Chandrasekhar
World Culture and the Black Experience,
by Ali A. Mazrui
Energy for Tomorrow, by Philip H. Abelson
Plato's Universe, by Gregory Vlastos
The Nature of Biography, by Robert Gittings
Darwinism and Human Affairs,
by Richard D. Alexander
Arms Control and SALT II, by W. K. H. Panofsky
Promethean Ethics: Living with Death,
Competition, and Triage,
by Garrett Hardin
Social Environment and Health, by Stewart Wolf
The Possible and the Actual, by François Jacob
Facing the Threat of Nuclear Weapons,
by Sidney D. Drell
Symmetries, Asymmetries, and the World of Particles,
by T. D. Lee

Symmetries, Asymmetries, and the World of Particles

T. D. Lee

UNIVERSITY OF WASHINGTON PRESS
Seattle and London

Printed in the United States of America

Library of Congress Cataloging in Publication Data

Lee, T. D., 1926–
Symmetries, asymmetries, and the world of particles.

(The Jessie and John Danz lectures)
Bibliography: p.
1. Symmetry (Physics) 2. Nuclear physics. 3. Particles (Nuclear physics) I. Title. II. Series.
QC174.17.S9L44 1987 530.1'42 87–23217
ISBN 0-295-96519-3

For Shanshi and Shanxuan

The Jessie and John Danz Lectures

In October 1961, Mr. John Danz, a Seattle pioneer, and his wife, Jessie Danz, made a substantial gift to the University of Washington to establish a perpetual fund to provide income to be used to bring to the University of Washington each year "distinguished scholars of national and international reputation who have concerned themselves with the impact of science and philosophy on man's perception of a rational universe." The fund established by Mr. and Mrs. Danz is now known as the Jessie and John Danz Fund, and the scholars brought to the University under its provisions are known as Jessie and John Danz Lecturers or Professors.

Mr. Danz wisely left to the Board of Regents of the University of Washington the identification of the special fields in science, philosophy, and other disciplines in which lectureships may be established. His major concern and interest were that the fund would enable the University of Washington to bring to the campus some of the truly great scholars and thinkers of the world.

Mr. Danz authorized the Regents to expend a portion of the income from the fund to purchase special collections of books, documents, and other scholarly materials needed to reinforce the effectiveness of the extraordinary lectureships and professorships. The terms of the gift also provided for the publication and dissemination, when this seems appropriate, of the lectures given by the Jessie and John Danz Lecturers.

Through this book, therefore, another Jessie and John Danz Lecturer speaks to the people and scholars of the world, as he has spoken to his audiences at the University of Washington and in the Pacific Northwest community.

Contents

Prologue

"Tell me, why should symmetry be of importance?" asked Chairman Mao Zedong.

That was on May 30, 1974, when China was still in the turmoil of the Cultural Revolution and the Gang of Four was at the zenith of its power. I was especially depressed to find, in that ancient land of civilization, that education had been almost totally suspended. I hoped desperately that somehow there would be a way to improve, however slightly, the course of events.

At about six o'clock that morning, the phone in my room at the Beijing Hotel had rung unexpectedly. I was told that Mao would like to see me in one hour at his residence in Zhong Nan Hai, inside the former imperial palace. I was even more surprised that when he saw me the first thing he wanted to find out about was symmetry in physics.

According to Webster's dictionary, symmetry means "balanced proportions" or "the beauty of form arising from such balanced proportions." In Chinese, symmetry is 对称, which carries an almost identical meaning. Thus it is essentially a static concept. In Mao's view, the entire evolution of human societies is based on dynamic change. Dynamics, not statics, is the only important element. Mao felt strongly that this also had to be true in nature. He was, therefore, quite puzzled that symmetry should be elevated to such an exalted place in physics.

During our meeting, I was the only guest. A small end table was placed between our chairs, on which there were pads, pencils, and the ever present green tea. I put a pencil on the pad and tipped the pad toward Mao and then back toward me. The pencil rolled one way and then the other. I pointed out that at no instant was the motion static, yet as a whole the dynamic process had a symmetry. The concept is by no means static; it is far more general than its common meaning indicates and is applicable to all natural phenomena from the creation of our universe to every microscopic subnuclear reaction. Mao appreciated the simple demonstration. He then asked more questions about the deeper meaning of symmetry, and also about other physics topics. He expressed regret that he had not had the time to study science, but he remembered a set of science books by J. Arthur Thomson which he had enjoyed reading when he was young.

Our conversation gradually shifted from natural phenomena to human activities. In the end, Mao accepted my limited proposal that the education of at least the very brilliant young students should be maintained, continued, and strengthened. This led, with the strong support of Zhou Enlai, to the elite "youth class," a special intensive education program for talented students from the early teens through college. It was established first at the University of Science and Technology in Anhui and later, because of its success, also at other Chinese universities.

The next day, at the airport, I received a farewell present from the Chairman: a four-volume set of the original 1922 edition of *The Outline of Science* by J. Arthur Thomson.

To the general chaos produced by the Cultural Revolution, this meeting brought only a minute amount of order. Nevertheless, in a very limited way perhaps it does indicate a correlation between man's intrinsic urge to search for the symmetry in nature and his desire for a society that is both meaningful and more balanced.

Symmetries
and
Asymmetries

JABBERWOCKY

'Twas brillig, and the slithy toves
Did gyre and gimble in the wabe:
All mimsy were the borogoves,
And the mome raths outgrabe.

She puzzled over this for some time, but at last a bright thought struck her. "Why, it's a looking-glass book, of course! And, if I hold it up to a glass, the words will all go the right way again."

—*Through the Looking Glass* by Lewis Carroll

Mirror Symmetry

Like Alice, physicists also know that there is a difference between right and left. But they also *did* believe that this difference is not an absolute one. If you look through a mirror, then right becomes left and left becomes right. While the looking-glass world is different, how can we be sure that our world is not in reality a looking glass of yet another looking-glass world? For a very long time it was regarded as a sacred principle that the laws of nature should be symmetrical under a mirror reflection (right-left symmetrical); that is, the world in the looking glass could be a real world.

In everyday life, right and left are obviously distinct from each other. Our hearts, for example, are usually not on the right side. The word *right* also means "correct," right? The word *sinister* in its Latin root means "left"; in Italian, "left" is *sinistra*. In English, one says "right-left," but in Chinese 左右: 左 (left) traditionally precedes 右 (right). Such asymmetry in daily life, however, is attributed either to the accidental asymmetry of our environment or to initial conditions. Before the discovery of right-left symmetry violation (parity nonconservation) at the end of 1956, it was taken for granted that the laws of nature are symmetrical under a right-left transformation.

Let me give you an example. Suppose there are two cars which are made exactly alike, except that one is the mirror image of the other, as shown in Figure 1. Car ɑ has a driver on the left front seat and the gas pedal near his right foot, while ɒ has the driver on the right front seat with the gas pedal near his left foot. Both cars are filled with the same amount of gasoline, which is assumed to have no impurities and is right-left symmetrical. Now, suppose the driver of Car ɑ starts the car by turning the ignition key clockwise and stepping on the gas pedal with his right foot, causing the car to move forward at a certain speed, say 30 mph. The other does exactly the same thing, except that he interchanges right with left; that is, he turns the ignition key counterclockwise and steps on the gas pedal with his left foot, but keeps the pedal at the same degree of inclination. What will the motion of Car ɒ be? The reader is encouraged to make a guess.

Probably your common sense will say that obviously both cars should move forward at exactly the same speed. If so, you are just like the pre-1956 physicists. It would seem reasonable that two arrangements, identical except that one is the mirror image of the other, should subsequently behave in exactly the same way in all respects, except of course for the original right-left difference. In other words, while right and left are different from each other it seems self-evident that, except for that difference, there should be no other difference. Therefore, which one we call "right" and which "left" would be entirely relative. This is precisely the right-left symmetry principle in physics.

Surprisingly, this is found to be not true. In 1956, C. S. Wu, E. Ambler, R. W. Hayward, D. D. Hoppes, and R. P. Hudson investigated the decay of polarized cobalt nuclei, Co^{60}, into electrons. Because these nuclei are polarized they rotate parallel to each other. The experiment consisted of two setups, identical except that the directions of rotation of the

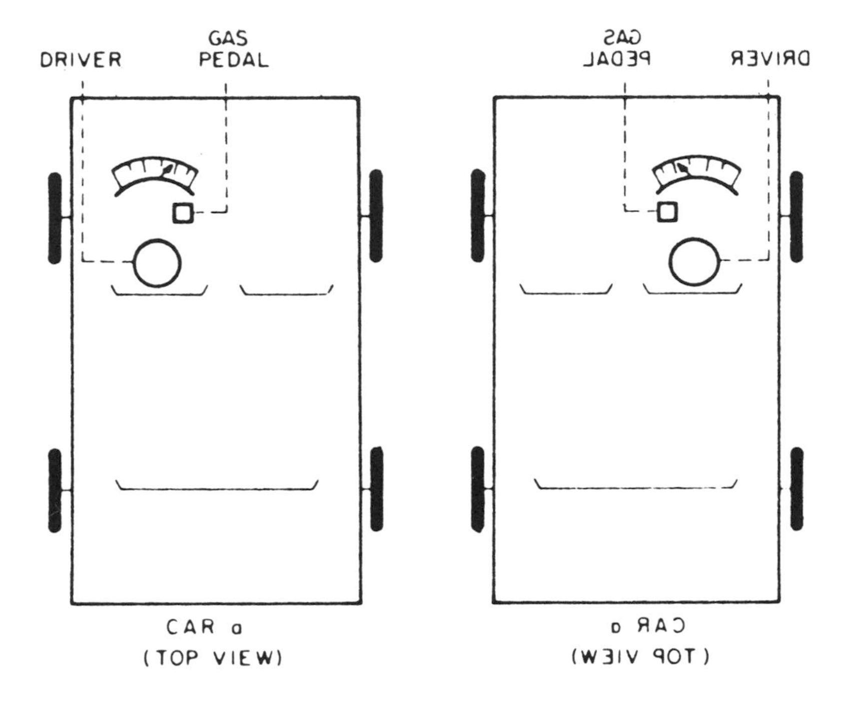

Figure 1. Two cars manufactured exactly alike except that one is the mirror image of the other.

initial nuclei were opposite; that is, each was a looking-glass image of the other. The experimenters found, however, that the patterns of the final electron distributions in these two setups are not mirror images of each other. In short, the initial states are mirror images, but the final configurations are not (see Figure 2).

Now come back to our example of two cars. In principle it is possible to install a β-decay source as part of the car's ignition mechanism. It would then be feasible, though perhaps not economical, to construct two cars that are mirror images of each other, but may nevertheless move in a completely different way: Car a moves forward at a certain speed, while car ɒ may move at a totally different speed, or may even move backward. That is the essence of the discovery of right-left asymmetry, or parity nonconservation.

Symmetries and Nonobservables

At this stage, it may be worthwhile to pause and indulge in some abstract thinking. When we say right-left symmetry, we imply that it is impossible to observe an absolute difference between right and left. (Of course one "knows" that right is distinct from left.) On the other hand, if it turns out than an absolute difference between them can be found, then we have right-left symmetry violation or right-left asymmetry.

Indeed, all symmetries are based on the assumption that it is impossible to observe certain basic quantities, which we shall call "nonobservables." Conversely, whenever a nonobservable becomes an observable, we have a symmetry violation. This will be the theme that runs through this entire essay.

To help you appreciate this line of reasoning, let me give another example. Consider the interaction energy V between two objects, say the earth and the sun, which we label 1 and 2. Imagine that a reference point has been arbitrarily chosen,

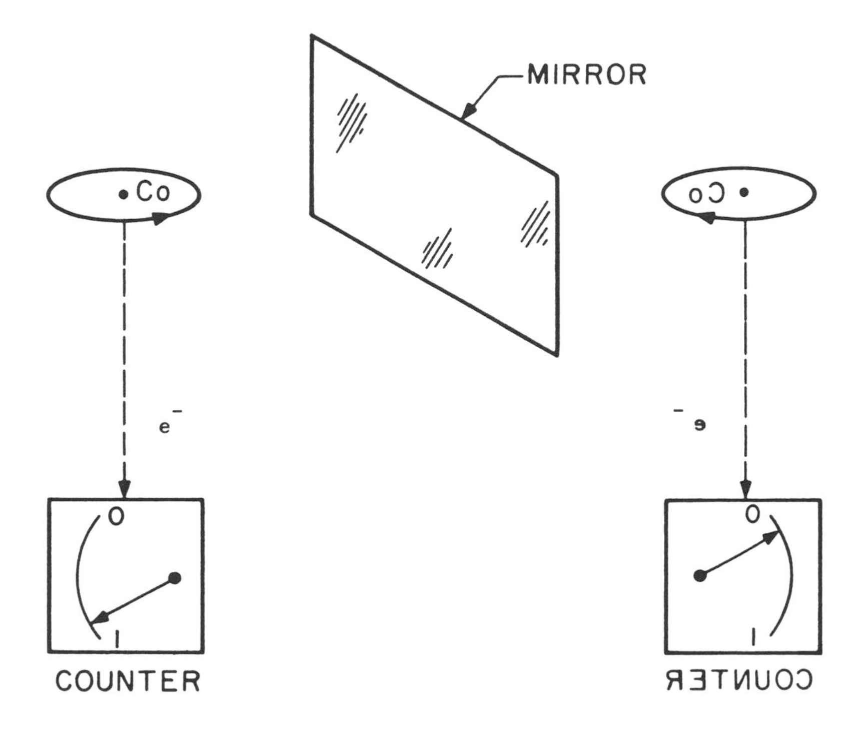

Figure 2. The initial setups of these two experiments on Co^{60} decay are exact mirror images, but the final electron distributions are not, as indicated by the different readings on the counters.

which we will call O. With respect to O, the positions of earth and sun can be represented by the vectors $\vec{r}_1$ and $\vec{r}_2$. Each vector denotes both a distance and a direction.* This setup is illustrated by the solid lines in Figure 3.

We now make the assumption that the absolute position of any object is a nonobservable. Of course the relative positions of any two objects to each other can be measured. Consequently, their interaction energy V should depend only on their relative positions; or, equivalently, V should be independent of the reference point O. Because V is independent of O, V must stay the same if we move the reference point O to O′ by a distance vector, say $\vec{\Delta}$. But, in the process, the position vectors of 1 and 2 with respect to the reference point are shown by the change from the solid lines to the dashed lines in Figure 3:

$$\vec{r}_1 \rightarrow \vec{r}_1 - \vec{\Delta} \quad \text{and} \quad \vec{r}_2 \rightarrow \vec{r}_2 - \vec{\Delta}, \qquad (1)$$

whereas V (being unchanged) must depend only on the difference $\vec{r}_1 - \vec{r}_2$ which is the relative position of 1 with respect to 2 (and remains the same when we move O to O′):

$$V = V(\vec{r}_1 - \vec{r}_2). \qquad (2)$$

Next, imagine that the position of the earth is moved by a small amount. The force on the earth is proportional to the corresponding rate of energy change. This same reasoning would also be applicable if the sun were moved instead. Let us now move the earth and the sun together by the same

*Readers who have not encountered vectors before may find these symbols compact and useful. For example, when one says "the distance between New York and Princeton is x miles," the x is just a number (since no direction is specified). But when one says "New York is x miles east of Princeton," there is a direction as well as the number x; together they form a vector $\vec{x}$, as indicated by the arrow sign on the top of $\vec{x}$.

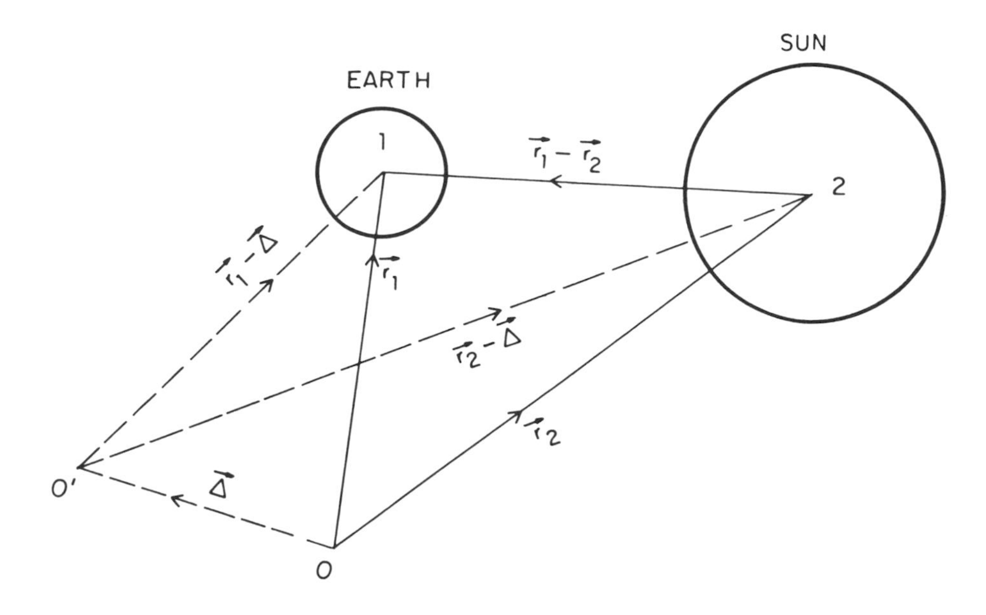

Figure 3. A translation of the reference point O to O′ does not change the relative distance vector between 1 and 2.

small amount. Clearly, this does not alter their relative positions $\vec{r}_1 - \vec{r}_2$. From Equation (2) we see that the energy V is also not changed. Since there is no energy change for the entire system, the total force must be zero; that is, the force acting on the earth cancels that on the sun. Therefore we have deduced Newton's Third Law (For every action there is an equal and opposite reaction) by the simple assumption that absolute position is a nonobservable.

Furthermore, because the rate of change of the momentum of any system is equal to the total force acting on that system, the absence of total force means that the total momentum is conserved (unchanged). The conservation law of momentum has now also been derived by our basic assumption that absolute position is a nonobservable.

If we review what we have just said, we will find that there are three distinct logical steps:

1. the physical assumption that absolute position is a nonobservable.
2. the implied property that the interaction energy V is unchanged under the connected mathematical transformation represented by Equation (1); and
3. the physical consequence of the momentum conservation law. Conversely, we may use the momentum conservation law to test whether absolute position is a nonobservable or not. In mathematical jargon, the transformation represented by Equation (1) is called a space translation. The property that the interaction energy V is unchanged is often referred to as "invariance" (in this case, the invariance of V under a space translation).

In an entirely similar way, we may assume absolute time to be a nonobservable. The physical law must then be symmetrical (invariant, i.e., not changing) under a time translation

$$t \rightarrow t + \tau$$

which results in the conservation of energy. (This is because the total energy does not change when we increase time t by an arbitrary amount τ; consequently the total energy must be a constant, independent of time.) By assuming the absolute spatial direction to be a nonobservable, we derive rotation symmetry and obtain the conservation law of angular momentum. (This derivation follows the same steps as when we derived momentum conservation under space translation; all that is needed is to replace "positions in space" by "angular positions" in Equation (1). Then space translation becomes angular rotation, and the implied momentum conservation becomes angular momentum conservation.)

This train of logic extends to all the symmetry principles used in physics, from relativity to quantum theory. It forms an extremely powerful tool in our theoretical analysis of nature. By starting from the very simple assumption of nonobservables, we can arrive at consequences which are far-reaching and universal, independent of the detailed structure of the particular system under consideration. There are few disciplines in any intellectual pursuit that can match the profound generality and the aesthetic simplicity of symmetry principles. (Further details are given in the Appendix.)

Asymmetries and Observables

Since nonobservables imply symmetry, any discovery of asymmetry must imply some observable. The experiment of Wu, Ambler, Hayward, Hoppes and Hudson, mentioned earlier during our discussion of mirror symmetry, also established the asymmetry between the positive and negative signs of electricity. In this connection, one may ask what exactly are the observables that have been discovered through these symmetry-breaking phenomena? We recall that in our usual terminology the sign of electric charge is merely a convention. The electron is considered to be negatively charged because we happened to assign the proton a positive charge, and the converse is also true. But now, with the discovery of asymmetry, is it possible to give an absolute definition to the sign of electric charge? Can we find some absolute difference between the positive and negative signs of electricity, or between left and right?

As an illustration, let us consider the example of two imaginary, advanced civilizations A and B, as in Figure 4. These two civilizations are assumed to be spatially completely separate from each other; nevertheless they manage to communicate, but only through electrically neutral and unpolarized messages, such as unpolarized light. After years

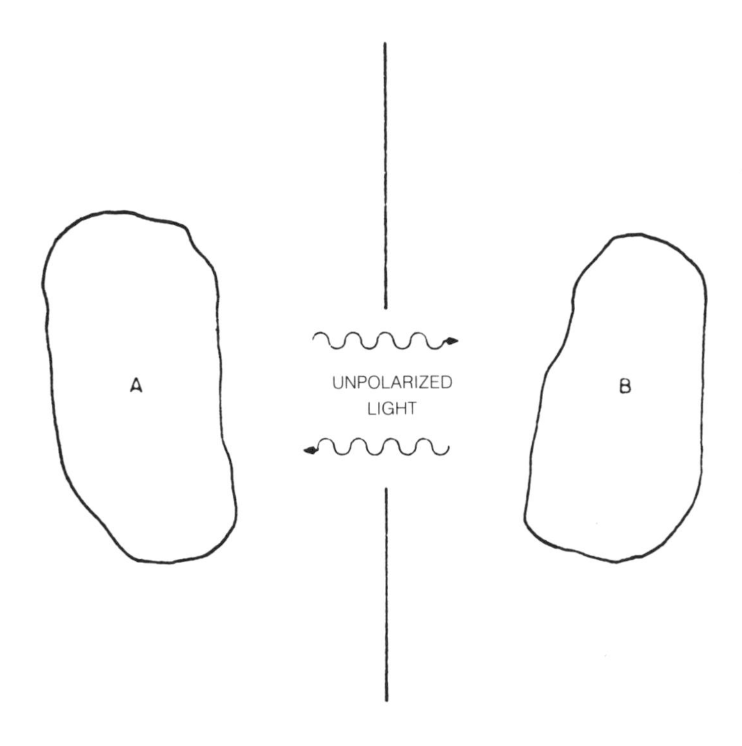

Figure 4. Two imaginary civilizations, A and B.

of such communication these two civilizations may decide to increase their contact. Being very advanced, they realize that they must first agree on (1) the sign of electric charge, and (2) the definition of a right-hand screw.

The first is important in order to establish whether the proton in civilization A corresponds to the proton or the antiproton in civilization B. Only if the protons in A are the same as those in B is a closer contact advisable. The definition of a right-hand screw is important if these two civilizations decide to have even closer contact, such as trading machinery. It is important that they should agree on the screw-convention before the actual arrival of the merchandise.

(The reader may wonder why civilization A couldn't simply send a "hand" directly to civilization B. This is not allowed, because if A is made of matter and B of antimatter, such a direct exchange may lead to annihilation and is therefore at least provocative, if not dangerous.) In any case, the academic problem that concerns us is whether it is possible to transmit both pieces of information by using only neutral and unpolarized messages. Without the discoveries of symmetry violations, made over the past three decades, this would not be feasible. Now, assuming that these two civilizations are as advanced as ours, such an agreement can in principle be achieved.

First, both civilizations should establish high-energy physics laboratories. (I am afraid that, in order to keep up with these two civilizations, the reader has to tolerate some technical high-energy physics terminology.) These laboratories can produce the long-lived neutral kaon $K_L{}^0$, which carries no electric charge, no electromagnetic moment of any kind, has no spin, is spherically symmetric and is about 1,000 times more massive than the electron. The $K_L{}^0$ is unstable and can decay into either an electron e^-, a neutrino* ν, and a positive pion π^+, or their antiparticles: a positron e^+, an antineutrino $\bar{\nu}$, and a negative pion π^-. These two different decay modes can easily be distinguished by a magnetic separation of e^+ from e^-. The physicists in these two civilizations would find that, although the parent particle $K_L{}^0$ is electrically neutral, nevertheless these two decay rates are different:

$$\frac{\text{rate } (K_L{}^0 \rightarrow e^+ + \pi^- + \nu)}{\text{rate } (K_L{}^0 \rightarrow e^- + \pi^+ + \bar{\nu})} = 1.00648 \pm 0.00035. \quad (3)$$

*A neutrino or its antiparticle, the antineutrino, is electrically neutral and has a spin but no mass; hence they always travel with the velocity of light c, and their energies are equal to c times their momenta. The charged pion is spherically symmetric, has no spin, has a mass about 280 times the electron's, and carries one unit of electric charge (positive for π^+, negative for π^-).

This is indeed remarkable since it means that by *rate counting* one can differentiate e^+ from e^-. Thus, there is an absolute difference between the opposite signs of electric charge. If we pause to think about this, we are struck by how truly extraordinary it is: the original particle, $K_L{}^0$, is completely neutral electrically. Therefore we would assume it should have no preference for positive or negative signs of electricity; yet it decays faster into e^+ than e^-! So, now, each civilization needs only to examine the faster decay mode and compare the charge of the final e with that of its "proton." If both civilizations have the same relative sign, then it means that they are made of the same matter.

We now come to the second task: the definition of a right-hand screw. This can be done by measuring the spin and momentum direction of the neutrino or antineutrino in the above kaon decay. Both the neutrino and the antineutrino possess a spin (angular momentum). For a neutrino, if one aligns one's left thumb parallel to its momentum, then the curling of one's four fingers would always be in the direction of its spin. Therefore, the spin and momentum direction of a neutrino defines a perfect left-hand screw, whereas the spin and momentum direction of an antineutrino defines a perfect right-hand screw. This property holds for neutrinos and antineutrinos everywhere, independently of how they are produced (see Figure 5).

Returning now to our two-civilization problem, we see that, by measuring the two $K_L{}^0$ decay rates and the screw sense defined by the neutrino, it is indeed possible for these two civilizations, through neutral unpolarized messages, to give an absolute definition of the plus and minus signs of electric charge, as well as of left and right.

The fact that we can give an absolute meaning to the sign of electric charge means that nature is not symmetrical with respect to that sign. In physicist's jargon, this is called "charge conjugation violation," or C violation. Likewise, the

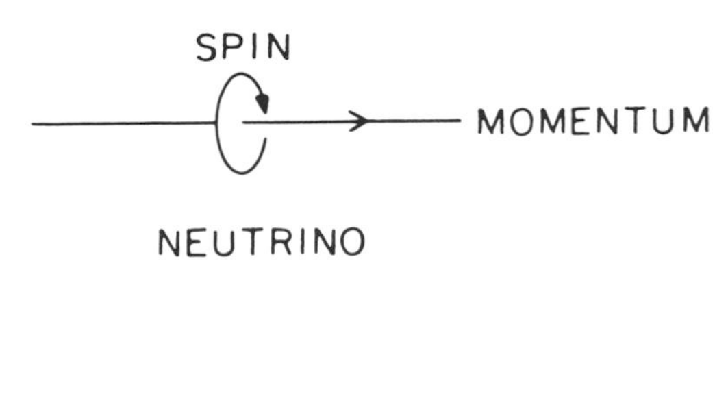

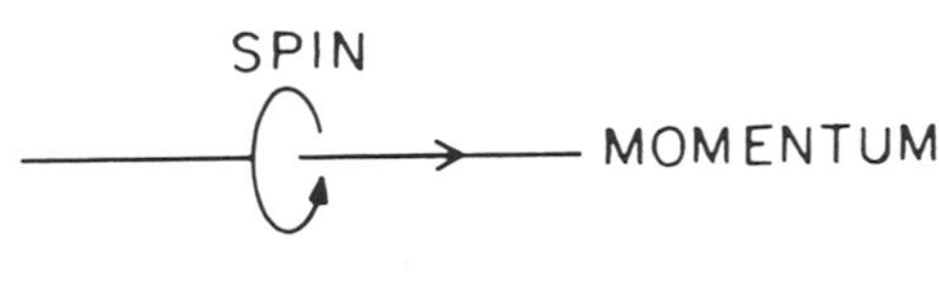

Figure 5. The spin momentum of a neutrino defines a left-hand screw; that of an antineutrino defines a right-hand screw.

fact that we are able to give an absolute definition of a right-hand screw implies right and left asymmetry, or asymmetry under mirror reflection. This is called "parity violation," or *P* violation. An interchange between positive and negative signs of electricity is represented by *C,* and an interchange between right and left by *P*.

It turns out that there is a close tie among three seemingly unrelated symmetries:

C sign change in electric charge,
P mirror reflection,
T time reversal.

As we have discussed, nature is not symmetrical under *C* or *P*. By more careful examination of kaon decay, it was found that nature is also not symmetrical under time reversal (which we will talk about in the next section). However, as far as we know, the combination of these three operations seems to be an exact symmetry. In other words, if we interchange

particle	$\leftrightarrows$	antiparticle,
right	$\leftrightarrows$	left,
past	$\leftrightarrows$	future,

all physical laws appear to be symmetrical and this is called *CPT* symmetry. From *CPT* symmetry, one can deduce that the masses of any particle and its antiparticle must be the same; their electric charges must be of the same magnitude and opposite signs.

Time Reversal

Time-reversal symmetry *T* means that the time-reversed sequence of any motion is also a possible motion. Some of you may think this absurd, since we are all getting older, never younger. So how can physicists even contemplate that the laws of nature should be time-reversal symmetrical?

In this sense we must distinguish between the evolution of a small system and a large system. Let me give an example. In Figure 6a each circle represents an airport, and a line indicates an air corridor. We assume that between any two of these airports the number of flights going both ways along any route is the same (this property will be referred to as microscopic reversability). Thus, a person in Wenatchee can travel to Seattle (Wenatchee is a city whose only air connection is to Seattle), then through Seattle to Vancouver or San Francisco (or New York). At any point of his travel, he can return to Wenatchee with the same ease. But suppose that in every airport we were to remove all the signs and flight information, while maintaining exactly the same number of flights, as shown in Figure 6b. A person starting from Wenatchee would still arrive in Seattle, since that is the only airport connected to Wenatchee. However, without the signs to guide him, it would be very difficult for him to pick out the return flight to Wenatchee from the many gates in the Seattle airport. The plane he gets on may be headed for San Francisco. If, in San Francisco, he then tries another plane again without any guidance, he could perhaps arrive in Tokyo. If he keeps on going this way, his chance of getting back to Wenatchee is very slim indeed. In this example, we see that microscopic reversibility is strictly maintained. When all the airport destination signs and other flight information are given clearly, then macroscopically we also have reversibility. On the other hand, if all such information is withheld, then the whole macroscopic process appears irreversible. Thus, macroscopic irreversibility is not in conflict with microscopic reversibility.

Time-reversal symmetry in physics refers to *microscopic reversibility* between all molecular, atomic, nuclear, and subnuclear reactions. Since none of these molecules, atoms, nuclei, and subnuclear particles can be easily marked, any macroscopic system in nature would exhibit irreversibility. This result is independent of microscopic reversibility. In any macroscopic process, we have to average over an immense number of unmarked microscopic units of atoms, molecules, and so forth (as in the example of unmarked airports and air routes), and that gives rise to *macroscopic irreversibility*. It is in this statistical sense of ever increasing disorder (entropy) that we define the direction of our macroscopic time flow. We may recall the words from *H.M.S. Pinafore*, by Gilbert and Sullivan,

> What never? No, never!
> What never? Well, hardly ever!

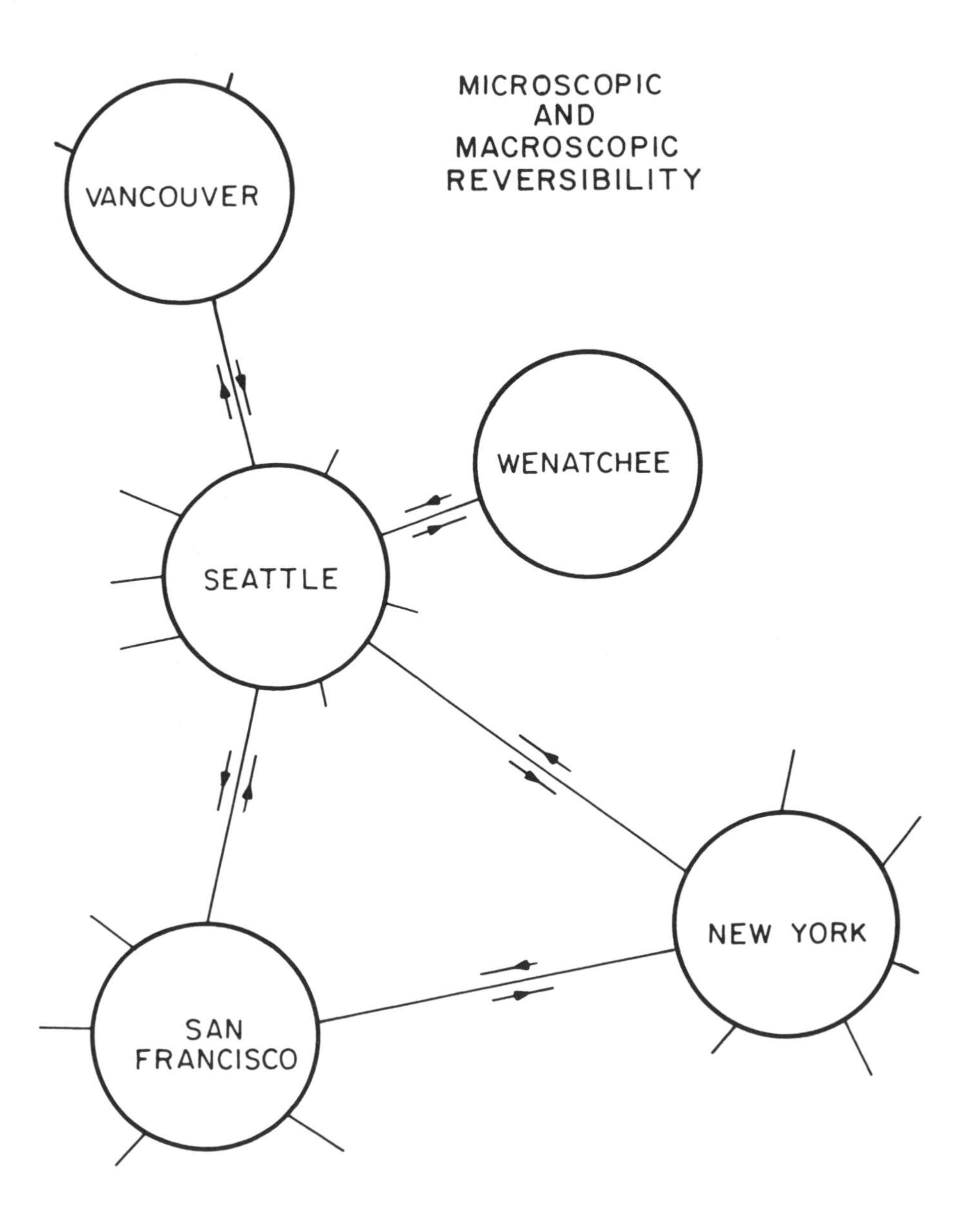

Figure 6a. In this example, microscopic reversibility means an equal number of flights in either direction on every air route. When the names of the airports, the numbers of the gates, and all flight information are known, there is also macroscopic reversibility.

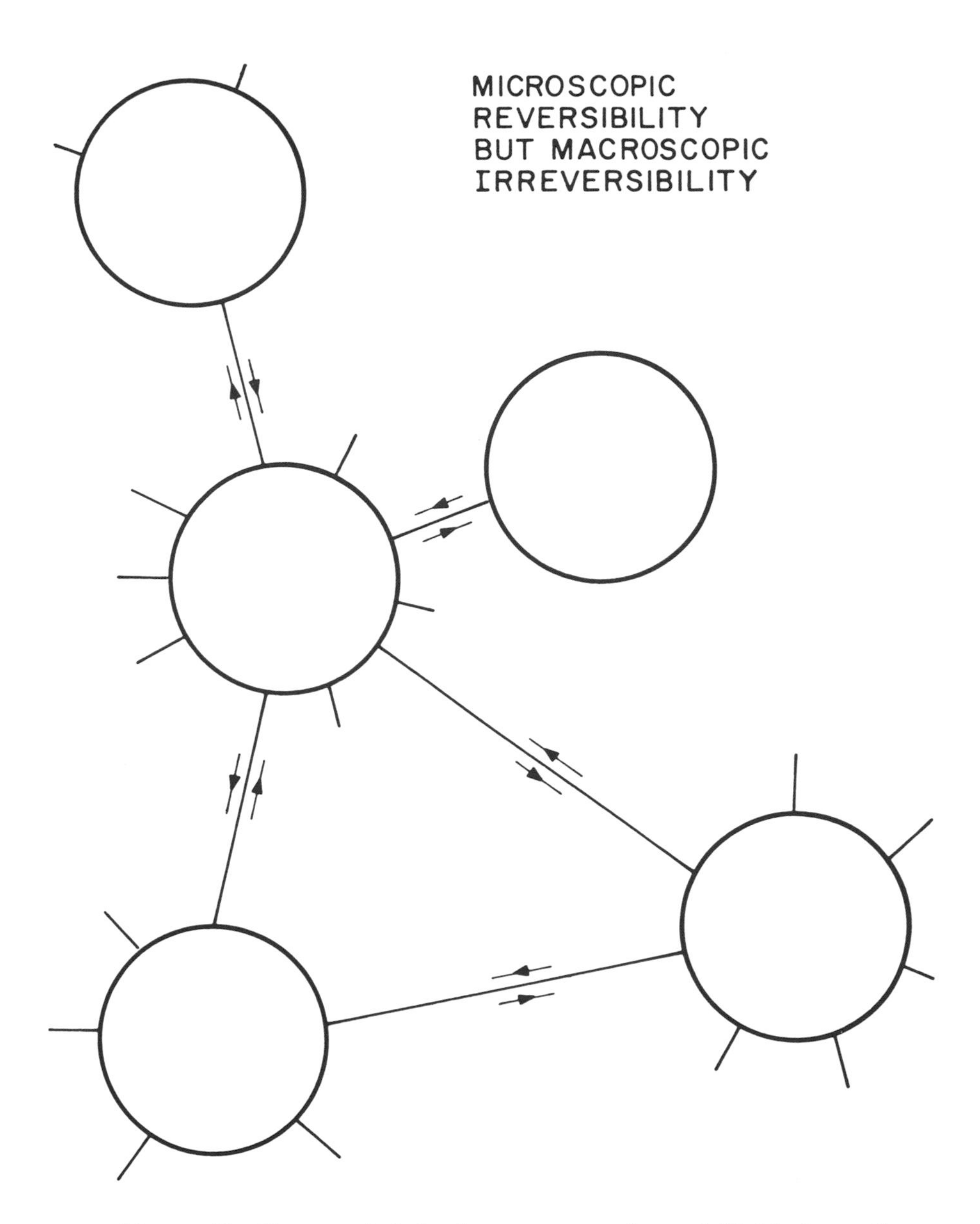

Figure 6b. If we maintain the same number of flights in each direction on any route (i.e., microscopic reversibility), but remove destination signs, gate numbers, and all other information, then it is nearly impossible to find our way back (i.e., macroscopic irreversibility).

The existence of the macroscopic time direction then leaves open the important question whether time-reversal symmetry, or microscopic reversibility, is true or not. Since 1964, after the discovery of *CP* violation by J. H. Christenson, J. W. Cronin, V. L. Fitch, and R. Turlay, through a series of remarkable experiments involving the kaons, it was found that the microscopic reversibility is indeed violated. Nature does not seem to respect time-reversal symmetry!

Asymmetrical Natural Laws, or Asymmetrical World?

In a sense, the discovery of "symmetry violations" is not that surprising. As we have discussed, the validity of all symmetry principles rests on the theoretical hypothesis of nonobservables. Some of these "nonobservables" may indeed be fundamental, but some may simply be due to the limitations of our present ability to measure things. As we improve our experimental techniques, our domain of observation naturally enlarges. It should not be completely unexpected that we may succeed in detecting one of those supposed "nonobservables" at some time, and therein lies the origin of symmetry breaking. When such a violation does occur, however, a deeper question is, how can we be sure that it is not the world at large that is asymmetrical? Is it possible that the fundamental laws of nature are nevertheless totally symmetrical?

What is the difference between these two views: an asymmetrical natural law? or an asymmetrical world? Insofar as we accept the fundamental law of nature to be immutable and permanent, while the world obviously undergoes continuous change, these two possibilities are clearly distinct from each other, though not mutually exclusive. An asymmetrical law implies an asymmetrical world, but not vice versa. Since we are perhaps more accustomed to a world which is somewhat skewed, it seems at least meaningful to inquire whether all the recent discoveries of symmetry violations are consistent with our fundamental physical laws being totally symmetrical.

Perhaps it should be emphasized that, in our previous discussions, we attribute all these asymmetries to the symmetry violation of physical laws. The reason is that all these asymmetrical reactions (β-decay, K-decay, and so forth) can occur in vacuum. Each decay seems to involve only an isolated system of a few particles. These asymmetry experiments can be repeated, and have been repeated, some thousands of times. Their outcomes are, by now, completely predictable; the resulting asymmetries do conform with great accuracy to a set of symmetry-violating physical laws, which we know precisely. In view of all this, one may wonder how it can be possible for anyone to contemplate seriously the opposite view: that the *fundamental laws of nature may nevertheless remain symmetrical.* In order to envisage such a seemingly implausible possibility, some additional new ideas must be introduced. We shall now go over the basis of what is often referred to (in the physics literature) as the "spontaneous symmetry-breaking" mechanism. In it, one assumes that the source of all asymmetries lies in the physical vacuum state.

Vacuum as a Physical Medium

What is a vacuum? We all know, for example, that the earth has an atmosphere. If we pump out all the air and all the matter, then what remains is the vacuum. But, insofar as we are incapable of switching off physical interactions, the vacuum could possess enormous complexity. As we shall see, virtual creation and annihilation of particle-antiparticle pairs can occur continuously in the vacuum state. Therefore, the vacuum resembles a physical medium.

In the last century, in order to understand how the

electromagnetic force, and later the electromagnetic wave, could be transmitted in space, the vacuum was viewed as a medium called aether. In his note 3075 on experimental research the famous British physicist, M. Faraday, wrote over a hundred years ago:

> For my own part, considering the relation of a vacuum to the magnetic force and the general character of magnetic phenomena external to the magnet, I am more inclined to the notion that in the transmission of the force there is such an action, external to the magnet, than that the effects are merely attraction and repulsion at a distance. Such an action may be a function of the aether; for it is not at all unlikely that, if there be an aether, it should have other uses than simply the conveyance of radiations.*

At that time, however, because Newtonian mechanics was the only available one, it was thought that aether would provide an absolute rest frame. Only in such a frame could one measure the true speed of light; in any moving frame, the speed of light would be modified because of the motion. As is well known, the fact that this turned out to be untrue caused the downfall of aether and the rise of relativity.

Because of the symmetry required by relativity, the speed of light measured in any moving frame is the same, independent of the velocity of the observer. Just by moving around, an observer cannot change the speed of light (relative to himself), nor can he excite the vacuum. Relativistic invariance is not everything, however; it does not imply that the vacuum is uncomplicated. Imagine that we are able to take a "picture" of the vacuum. If we take a very, very long exposure, then the vacuum might appear to be deceptively simple, as shown in Figure 7a.

*Michael Faraday, *Experimental Researches in Electricity,* 3 vols. (London: R. and J. E. Taylor, 1839–55), 3:330–31.

Figure 7a. A "picture" of the vacuum if the exposure time were infinite. All fluctuations would then cancel out, and the vacuum appear quiescent.

Although the vacuum does not contain any matter, it still retains all physical interactions. Therefore, the vacuum would "look like" Figure 7b if the exposure time were short enough to catch the fluctuations due to the virtual excitations produced by interactions. These fluctuations would average

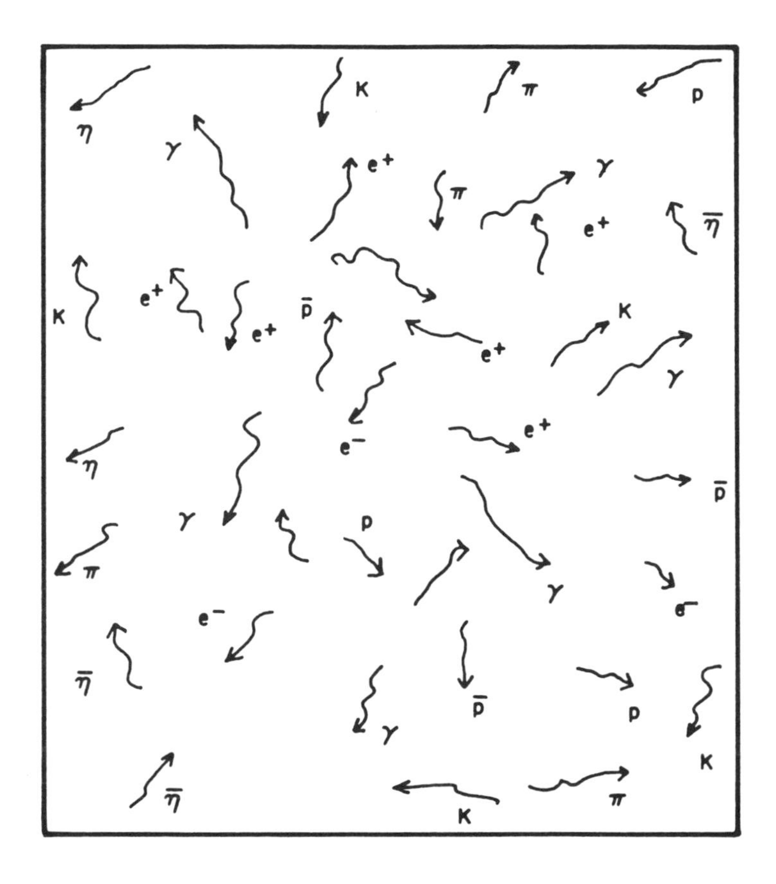

Figure 7b. If the exposure time were short, then one could see all the fluctuations, and the vacuum would appear complicated.

out to zero over a sufficiently long time interval, and that is why Figures 7a and 7b look so different. Because of the complexity of the vacuum, it becomes conceivable that, like any physical medium, the vacuum may appear asymmetrical.

Missing Symmetry and Spontaneous Symmetry-Breaking

As explained in the section on symmetries and observables, each symmetry results in a conservation law: right-left symmetry implies parity conservation, abelian gauge symmetry* gives charge or hypercharge conservation, and nonabelian gauge symmetry leads to isospin conservation. If however, we add up all these supposedly conserved quantities (called symmetry quantum numbers), such as parity, hypercharge, isospin, . . . , of all matter, we find these numbers to be constantly changing and therefore not conserved:

$$\text{change of} \left\{ \begin{array}{l} \text{parity} \\ \text{hypercharge} \\ \text{isospin} \\ \quad\vdots \end{array} \right\}_{\text{matter}} \neq 0. \qquad (4)$$

Aesthetically, this may appear disturbing. Why should nature abandon perfect symmetry? Physically, this also seems mysterious. What happens to these missing quantum numbers? Where do they go? In the spontaneous symmetry-breaking mechanism mentioned earlier, we assume that matter alone does not form a closed system. The hypothesis is that symmetry can be restored if we also include the vacuum. In other words, Equation (4) is replaced by:

$$\text{change of} \left\{ \begin{array}{l} \text{parity} \\ \text{hypercharge} \\ \text{isospin} \\ \quad\vdots \end{array} \right\}_{\text{matter + vacuum}} = 0. \qquad (5)$$

*See Appendix for definitions.

As a bookkeeping device, this is clearly possible. Unless we have other links connecting matter with vacuum, however, how can we be sure that this idea is right, and not merely a tautology?

The situation is somewhat similar to the deduction of retirement contributions from one's monthly paycheck. How can one know that the deduction is really for one's retirement? In all likelihood, it will not be possible to get the full value back in any case. If, however, when the time comes for retirement one does not get anything back, then one may question the reliability of the accounting. The same is true for the missing symmetry. We can always assume that whatever amount of symmetry is missing from matter has gone into the vacuum. But the key problem is whether it is possible to change the vacuum so that some of the missing symmetry may return to matter. If the vacuum indeed behaves like a physical medium, then it must be possible to change its properties by varying external conditions. This approach may lead to some crucial tests of whether symmetry principles can be maintained in the sense of Equation (5).

Relativistic Heavy Ion Collisions (RHIC)

Since the vacuum permeates the universe, it is virtually impossible (in the context of human endeavor) to change the entire vacuum. The extent of most elementary particles, however, is only about 10^{-13} cm (a fermi). If we can excite the vacuum over a region of linear size much greater than a fermi by putting in high energy, then so far as the physical properties of the particles within the region are concerned, it would be almost as if the whole vacuum were changed. A most effective way to precipitate such a change is to use relativistic heavy ion collisions. We take advantage of the fact that the diameter of a typical heavy ion nucleus can be about 10 fermis. By accelerating two beams of, say, U (uranium)

Figure 8. The Relativistic Heavy Ion Collider (RHIC) at Brookhaven National Laboratory will use the heavy ions from the Alternating Gradient Synchroton (AGS) (photo courtesy of Brookhaven National Laboratory)

nuclei at very high energy and making a head-on collision, we can heat up the colliding nuclei which, in turn, change the properties of the underlying vacuum. The changed vacuum will manifest itself as a "bubble" for a short time; occasionally it may survive briefly even after the colliding nuclei have flown apart. In principle, we may then test the physical laws in a different vacuum and thereby verify some of our theoretical ideas.

Figure 8 shows the overall picture of the RHIC project at Brookhaven National Laboratory. At present the three-and-a-half-kilometer-long RHIC tunnel is already dug, but the machine is still in the R and D (research and development) stage. The idea is to produce head-on collisions between U nuclei at 100 GeV* per nucleon (that makes a total of about 30 trillion electron volts per U nucleus). Through these collisions we can study, on the one hand, possible new forms of matter created through the fusion of the two nuclei; they can also enable us to examine the properties of the background vacuum after the colliding nuclei have passed. None of these experiments will be easy. But if it turns out that the vacuum indeed behaves like a physical medium, and if through physical means we can really alter the properties of the vacuum, then the microscopic world would become inextricably connected to the macroscopic world† through the ever present vacuum. It is likely that, by exploring the properties of the vacuum, we may be led to discoveries far more exciting than those that we have encountered so far.

*GeV stands for giga electron volts, 10^9 eV.

†The macroscopic world is the one we are familiar with; it consists of stars, planets, people, and other large-scale objects. The microscopic world refers to the world of particles, such as electrons, neutrinos, photons, protons and other small particles.

The World of Particles

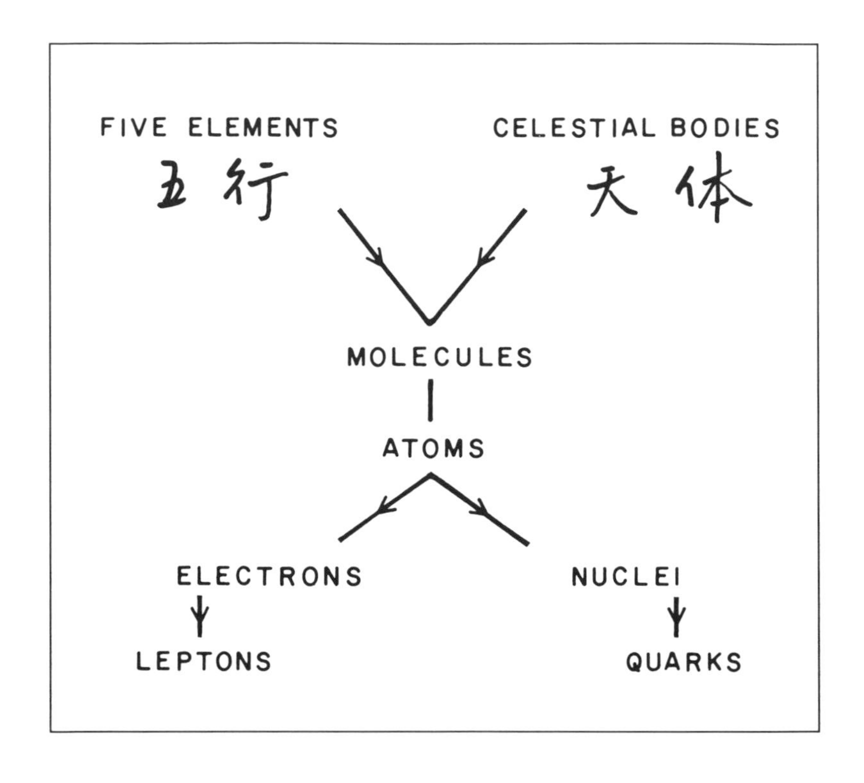

Figure 9. Evolution of "particle physics" over the centuries.

Particles and Accelerators

The origin of "particle physics" is tied to the assumption that the physical world is made up of smaller units. These units are called "particles." Throughout the history of physics, most of these units have invariably turned out to be themselves composed of even smaller ones. Consequently, there cannot be a permanent meaning to the term "elementary particle" or to the discipline "particle physics."

In ancient Chinese civilization, the world was thought to be composed of five elements:

金 gold,
木 wood,
水 water,
火 fire, and
土 earth.

An ancient Chinese particle physicist could therefore be any scholar interested in these five elements. The elements played a dual role; they were also thought to represent the planets:

Venus (gold)—metal,
Jupiter (wood)—organic,
Mercury (water)—liquid,
Mars (fire)—gas,

and Earth is self-explanatory. The bond between physics, astronomy, and cosmology is, as one can see, an old tradition.

Gradually, with the advance of astronomy it seemed reasonable to regard the stars as the basic units of the whole universe. The study of celestial bodies would then be the main occupation of "particle physicists." Indeed, that was precisely what Galileo and Newton did. Their work led to what we have today. In the nineteenth century the basic units underwent more changes, evolving eventually into molecules and atoms. Later, atoms resolved into even smaller units: electrons and nuclei, and finally into the leptons and quarks of today.

We all know that the basis of physics is experiment. Without it, physics would be reduced to mere philosophic speculation. The chief instrument for doing modern particle physics is the accelerator. These accelerators come in two basic forms, circular and linear. Both are important, and both are quite costly. Without such appropriate instruments, however, it is not possible to do high-energy physics experiments. It may be of interest to give a brief review of the history of instruments used in physics.

The basic ingredient of any successful experiment is the proper equipment. That this has been true ever since the beginning of physics can be illustrated by Figure 10. If Archimedes had not had this wonderful circular instrument (his tub), it might have been difficult for him to formulate his famous principle.

Even in ancient times, linear instruments also played an equally important part. Figure 11 is a picture of the original telescope used by Galileo. Without this marvelous linear tool, it would have been impossible for Galileo to do some of the particle physics experiments of his day. And, if we did not have those, we certainly would not have the particle physics of today.

As the basic units evolved, however, from celestial bodies to molecules and atoms, then to electrons and nuclei, their sizes diminished rapidly. On the other hand, the instruments used by physicists got larger and larger. This is indeed quite remarkable, as we have already seen from the size of the

ARCHIMEDES erſter erfinder ſcharpffſinniger vergleichung/
Wag vnd Gewicht/durch außfluß des Waſſers.

Figure 10. A sixteenth-century wood engraving of Archimedes (Bibliothèque Nationale, Paris; photo: Regal).

Figure 11. The telescope used by Galileo, now in the Museum of Physics in Florence (photo: Alinari).

RHIC project at Brookhaven. Figure 12 is a picture of the accelerators at SLAC (Stanford Linear Accelerator Center). The straight line at the top that crosses the highway is the two-mile-long linear accelerator, and the dashed line that curves around the center is the new Stanford Linear Collider (SLC) which will produce head-on collisions between electrons and positrons at 60 GeV each. Figure 13 shows the one-kilometer-radius Tevatron at Fermilab, south of Chicago, which is capable of colliding a one-TeV (10^{12} eV) proton on a one-TeV antiproton.

RHIC, SLC, and Tevatron are three examples of the types of superaccelerators in particle physics that are either under consideration or near completion in the United States. There are also important superaccelerator projects in Europe, Japan, and the Soviet Union, such as LEP at CERN (Centre Européen pour la Recherche Nucléaire) in Geneva, HERA at DESY (Deutsches-Electronen-Synchrotron) in Hamburg, TRISTAN at KEK (Kō Energy Kenkyūsho, which means High Energy Institute) in Tsukuba, Japan, and IHEP Accelerating Storage Complex at Serpukhov in the USSR. The biggest accelerator project of all is the recently approved Superconducting Supercollider (SSC) which will have a ring 52 miles in circumference (versus the next largest, the 16-mile circumference LEP project at CERN).

Why should these accelerators be so large if their purpose is only to explore the ultrasmall region occupied by the subnuclear particles? The reason is the uncertainty principle formulated by W. Heisenberg in 1925. It states that, for any experiment, there is an intrinsic inequality:

$$\Delta p \cdot \Delta x \geq \frac{\hbar}{2}$$

Figure 12. The Stanford Linear Accelerator (SLAC) (photo courtesy of SLAC).

Figure 13. Fermilab (photo courtesy of Fermilab).

where $\hbar$ is Planck's constant divided by 2π, and Δx, Δp refer to the uncertainties in distance and momentum determination in the same experiment. Consequently, in order to study physics at a small distance, say Δx, the energy E required must be larger than the minimal momentum uncertainty Δp times the velocity of light c:

$$E > \Delta p \cdot c \geq \frac{1}{2} \frac{\hbar c}{\Delta x}$$

For example, in order to study the structure of an electron, since its size* is about 10^{-11} cm, the energy required must be bigger than 1 MeV (10^6 eV). Because the proton is smaller than the electron by a factor of a thousand, in order to study proton structure the energy must be higher than 1 GeV. Since the size of the newly discovered intermediate boson is even smaller, about 10^{-16} cm, intermediate boson physics requires an energy of about 100 GeV or higher, which is the energy range of SLC and the Tevatron collider. Because machine size grows with energy, we must have larger and larger accelerators.

Discoveries and the Laws of Physicists

When each new accelerator is proposed, theorists are employed like high priests to justify and to bless such costly ventures. Therefore it pays to look at the track record of theorists in the past, to see how good their predictions were before experimental results.

Figure 14 lists almost all the major discoveries made in particle physics for more than three decades. It is of interest to note that, with the exception of the antinucleon ($\bar{p}$ and $\bar{n}$) and the intermediate bosons ($W^{\pm}$ and Z^0), *none* of these

*Here "size" refers to what is technically called the Compton wavelength, after the American physicist A. H. Compton.

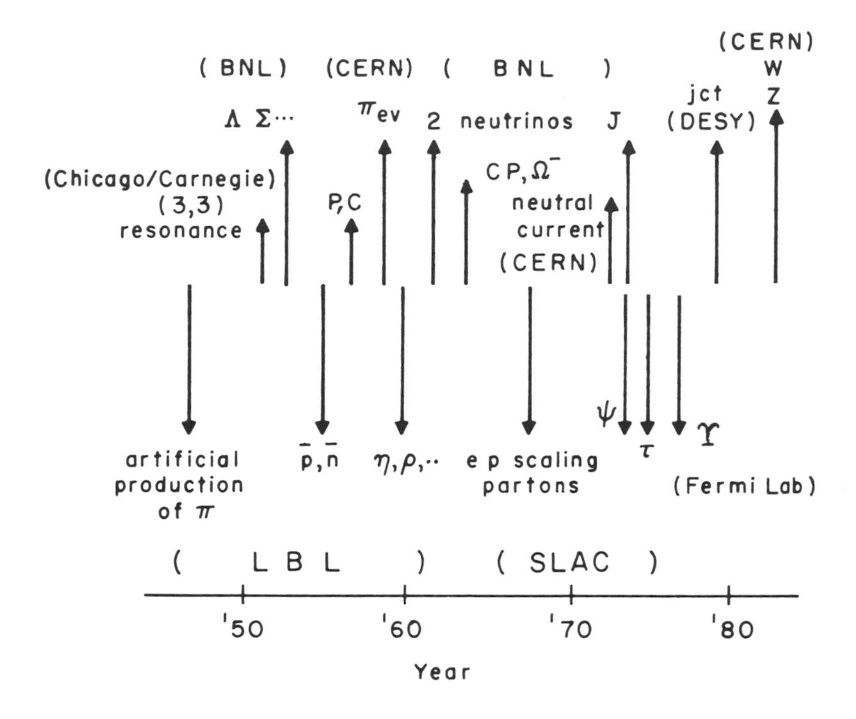

Figure 14. Major discoveries in particle physics.

Each line indicates a major achievement; its position gives the approximate time. Each arrow points to the name of a discovery and the laboratory responsible. Starting from the left, we have the artificial production of π made at Lawrence Berkeley Laboratory (LBL) in 1947–48, the discovery of the (3,3) resonance in 1953 at the University of Chicago and the Carnegie Institute of Technology (now Carnegie-Mellon University), the investigation of the dynamics of Λ, Σ, and other particles at BNL in 1953–54, the discovery of the antiproton and antineutron in 1955 at LBL, the discovery of P and C violations in 1956–57 (at Columbia and other institutions), the discovery of $\pi \rightarrow e\nu$ at CERN in 1958, and so on.

landmark discoveries was the original reason given for the construction of the relevant accelerator.

Let us start with the first discovery on the chart. When Ernest Lawrence built his 184-inch cyclotron, the energy was thought to be below pion production. Therefore, after the cyclotron was turned on, even though pions were produced abundantly, for a long time nobody noticed them. They were finally discovered accidentally and turned out to be the most important contribution made by that cyclotron.

The progress of particle physics is closely tied to the discovery of resonances,* which started with the (3,3) level first produced at the Chicago cyclotron (the second item in Figure 14). Yet even the great Enrico Fermi, when he proposed the machine, did not envisage this at all. After its unexpected discovery, for almost a year Fermi expressed doubts whether it was a genuine resonance. A similar story can be told about the next landmark discovery. When the Cosmotron was constructed at Brookhaven, some of the leading theorists thought that the most important high-energy problem was to understand the angular distribution of proton-proton collisions, which remains mysteriously flat even at a few hundred MeV, although at that energy the dynamics of the collision is quite complicated. Many different levels (*s, p, d, f, g*, . . .) are all involved. Why should they conspire to make a flat angular distribution? But, as it turned out, when the energy increases the angular distribution of proton-proton collisions no longer remains flat and becomes quite uninteresting. Instead, it was the production and decay dynamics of the strange particles, Σ, Λ, . . . that put the Cosmotron on the map.

*A resonance in particle physics means simply a fixed energy level. The word was carried over from the tuning fork, which vibrates (resonates) only at certain frequencies, similar to the radiation emitted during the transmutation of particles from one energy level to another.

We could go on and on, and the same pattern would repeat itself throughout this list. This leads to my *first law of physicists:**

Without experimentalists, theorists tend to drift.

There is no reason for us to believe that it will change, nor should we expect too much from our present theorists for the prediction of the future.

Look at Figure 14. You will notice that the density of great discoveries per unit time is quite uniform and averages out to about one in two years. Let us hope that this long-standing record of constant rate of discovery can be maintained. In order to achieve that, we must have good experiments.

We now come to my *second law of physicists:*

Without theorists, experimentalists tend to falter.

An example is the search for "neutral currents" in weak interactions. When high-energy neutrino experiments were proposed in 1960, it was suggested that perhaps this could be a tool to uncover neutral currents.† After two neutrinos were found in 1962, intensive experimental efforts were made to search for such a current. At that time, however, there was no theoretical guidance as to its magnitude. A year later, in 1963,

*To understand the laws of physicists, it is not at all necessary to know any law of physics.

†In a typical weak interaction, a neutron may beta-decay to become a proton (giving off an electron and an antineutrino). Conversely, a neutrino can convert a neutron into a proton plus an electron. Because there is a change of electric charge between neutron and proton, such reactions are called charge-current processes. Neutral current refers to a new kind of weak interaction in which a neutrino simply collides with a neutron without converting it into a proton.

an upper limit was set for the ratio of neutrino events due to neutral vs. charged currents (Proceedings of the Siena Conference):

$$\frac{\text{neutral current events}}{\text{charged current events}} < 3 \times 10^{-2}. \qquad (1963)$$

A decade later, however, in 1974, after the theoretical progress made by S. Weinberg and others, new experiments were performed. The same ratio turned out to be much larger:

$$\frac{\text{neutral current events}}{\text{charged current events}} = .42 \pm .08 \qquad (1974)$$

in good agreement with the theoretical model. The reason for the large discrepancy between these two results was never explained.

Another good example is the history of the "Michel parameter" in μ-decay. The momentum of the final e in μ-decay

$$\mu \rightarrow e + \nu_\mu + \bar{\nu}_e$$

varies from 0 to its maximum value. The electron distribution can be plotted against

$$x = (\text{momentum of } e) \,/\, (\text{its maximum value}),$$

and is characterized by the well-known Michel parameter, ρ, which can be any real number between zero and one. The

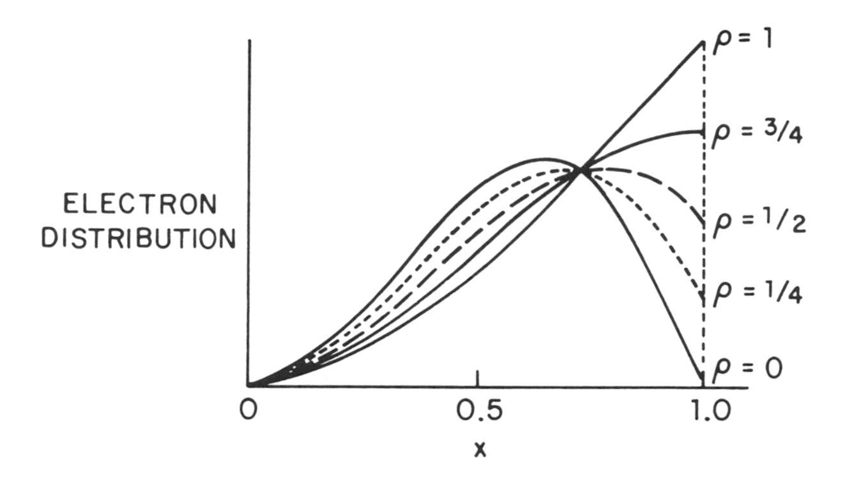

Figure 15. The distribution of electron energy in μ-decay.

Michel parameter measures the height of the endpoint of the electron distribution at the maximum electron momentum $x = 1$, as is shown in Figure 15. We see that different ρ-values give quite different electron distribution curves.

The Michel parameter has been under intensive experimental investigation since 1949. It is instructive to plot the experimental value of ρ against the year when the measurement was made. As shown in Figure 16, historically it was first found that ρ was near 0. Subsequent experiments, however, yielded different values; ρ slowly drifted upward. Only after parity nonconservation in 1957, when the theorists were able to make a precise prediction, $\rho = 3/4$, did the experimental value also begin to converge, finally reaching excellent agreement with theory in the sixties. When one looks at Figure 16, one is struck by the remarkable fact that at no time did the "new" experimental value lie outside the error bars of the preceding one.

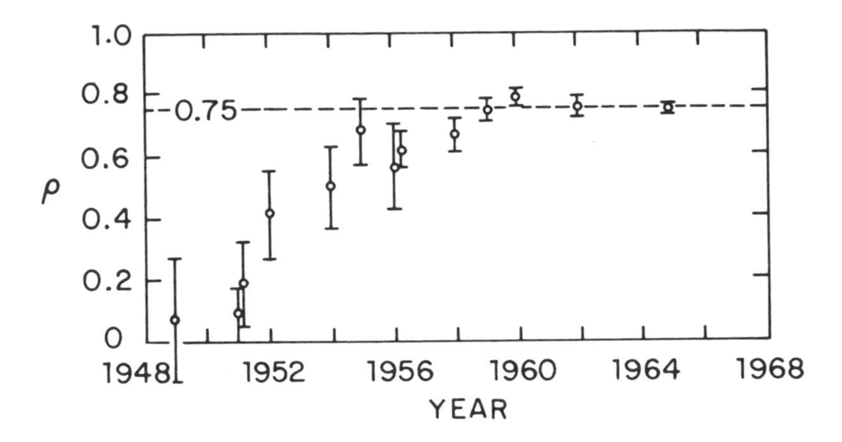

Figure 16. The change of the Michel parameter ρ from year to year.

I hope these examples explain the two laws of physicists and illustrate the interdependence of theory and experiment.

Present Status

The close collaborations between theorists and experimentalists over the past three decades have led to our present status, which is summarized in Table 1. There are three general classes of interactions: strong, electroweak, and gravitational. The strong interaction describes the forces that form protons and neutrons and combine them into various nuclei; the basic building blocks are the quarks. We think there are three families of quarks, with each family made up of two members. They are

up (u), down (d),
charm (c), strange (s), and
top (t), bottom (b).

Table 1. Present Status of Particle Physics

Interaction	Participating Particles	Carrier	Theory
strong	quarks: u, d, c, s, (t), b	gluon	quantum chromodynamics (*QCD*)
electroweak	quarks (listed above) and leptons: e, ν_e, μ, ν_μ, τ, ν_τ	photon, and intermediate bosons: $W^\pm$, Z^0	$SU(2) \times U(1)$ (including quantum electrodynamics [QED])
gravity	everyone and everything	graviton	general relativity

Five of them have been discovered, and only one, the top (t), is pending. The well-known protons and neutrons are formed of three quarks, and most of the mesons are made of a quark-antiquark pair. The strong interaction is carried from one place to another by the gluons, and the theory describing it is called quantum chromodynamics.

The electromagnetic and the weak interactions are now unified into a single class, called electroweak. The participating particles are quarks and leptons. The leptons also consist of three families. Like the quarks, each family also has two members. They are

electron (e), e-neutrino (ν_e),
muon (μ), μ-neutrino (ν_μ), and
tau (τ), τ-neutrino (ν_τ).

The electroweak interactions are carried by the photon and the recently discovered intermediate bosons $W^{\pm}$, Z^0. The theory describing the electroweak interaction is called $SU(2) \times U(1)$ gauge theory, more commonly known as the standard model. An integral component of the standard model is quantum electrodynamics, which describes the electromagnetic interaction.

The gravitational force is carried by the graviton, and the gravitational interaction is participated in by everyone and everything, including the graviton. The theory is the well-accepted general relativity of Einstein.

Considering that two decades ago the number of elementary particles responsible for the strong interaction consisted of the proton, the neutron, three pions, four kaons, three rho mesons, the baryon decuplets (consisting of ten different member particles), and an assorted army of many, many others, this is indeed a great simplification. What is more impressive is that the guiding principles for writing down these theories are extremely simple, based almost entirely upon symmetry principles that we discussed in the preceding chapter.

After saying these good things about our achievement, one may ask "Is there any problem?" This, then, leads to my next topic.

Two Puzzles

The two major puzzles that face us are (1) missing symmetries and (2) unseen quarks.

As we have said, symmetry implies conservation. Since our entire edifice of interactions is built on symmetry assumptions, there should be as a result a large number of conservation laws. The only trouble is that almost all of these conservation laws have been violated experimentally. This is the essence of the first puzzle, missing symmetry, which has been discussed before. As I mentioned in the preceding chapter, this difficulty could be resolved by introducing a new element, the vacuum. Instead of saying that the symmetry of all matter is being violated, we suggest that all conservation laws must take both matter and vacuum into account. If we include matter together with vacuum, then an overall symmetry could be restored.

We now turn to the second puzzle, unseen quarks.

Quarks

Quarks and antiquarks are supposed to be the constituents of all strongly interacting particles: protons, neutrons, pions, kaons, and so forth.

A pion is made of a quark and antiquark pair. Therefore, one would expect that, if a pion is split, what would come out should be a quark and an antiquark. It is quite easy to split pions in high-energy accelerators. The strange thing is that what emerges is never a quark and an antiquark, but instead more pions and other mesons.

A proton is made of three quarks. Again, it is not difficult to split a proton open in a high-energy collision, but what comes out of the collision are never free quarks, just nucleons (protons and neutrons), antinucleons, pions, and other mesons.

This is somewhat strange. But we may explain this by using the analogy of the magnet. A magnet has two poles, north and south. Yet, if you break a bar magnet in two, each becomes a complete magnet with two poles. By splitting a magnet open you will never find a single pole (magnetic monopole). This is illustrated in Figure 17. In our usual description, a magnetic monopole can be considered as either a fictitious object (and therefore unseeable) or a real object but with exceedingly heavy mass beyond our present energy range (and therefore not yet seen). In the case of quarks, as I will

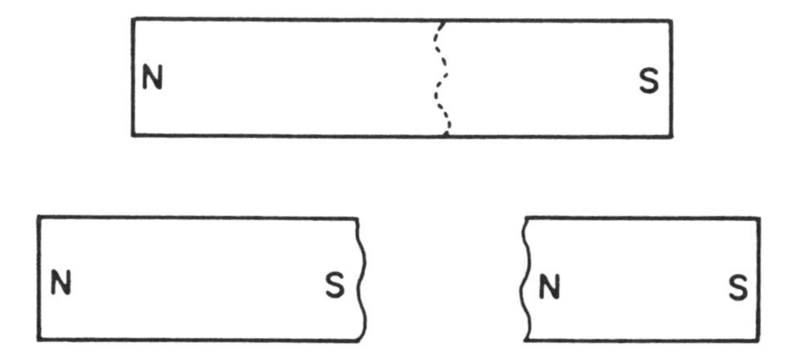

Figure 17. You cannot set a magnetic pole free. Every magnet has one north pole and one south pole. When you break it, you create more magnets, but never free poles.

explain, there are very good reasons to believe them to be physically *real and with very light masses*. If so, why don't we ever see free quarks? This is, then, the real puzzle.

In fact, quarks have many other peculiar properties. Because three quarks make a proton, the elementary electric charge of a quark is therefore a third of that of the proton. Although no one has ever seen a free quark, both the mass and the electric charge of quarks have already been determined experimentally. These are given in Table 2 together with some of their other known properties. Before discussing the experimental supporting evidence, we note that the first row lists the quark masses. The light ones are only about 5 or 10 MeV, which is around 1/200th or 1/100th of a proton mass. The second row gives their electric charges, which are −1/3 or 2/3 of a proton's. The third row tells us that each quark has three species, commonly referred to as three different "colors"; that is, there are three different species of up quarks, three different down quarks, and so forth. This characteristic was first introduced theoretically because of Pauli's exclusion principle, which states that no two identical

Table 2. The Names and Properties of the Quarks

Quark	u	d	s	c	b	t
mass	~10 MeV	5 MeV	100 MeV	2 GeV	5 GeV	?
charge/e Q_q	$\frac{2}{3}$	$-\frac{1}{3}$	$-\frac{1}{3}$	$\frac{2}{3}$	$-\frac{1}{3}$	$\frac{2}{3}$
species (color)	3	3	3	3	3	3
interaction strength	strong at ~1 GeV energy, but becomes 0 as energy approaches ∞					

particles of 1/2 spin (such as nucleons, electrons, and neutrinos) can be put into the same orbit.* Because three quarks make up a half-spin nucleon, the quarks must themselves have half spins (with two quark spins parallel to the net nucleon spin and the other quark spin antiparallel). Consequently, quarks must also obey Pauli's exclusion principle. In order to put three quarks into the same orbit to make a nucleon (proton p or neutron n), and yet not violate the Pauli principle, it is hypothesized that each quark actually consists of three different species. This was later verified experimentally.

Another interesting property is shown in the last row in Table 2. The interaction strength between quarks, while strong at around 1 GeV, becomes very weak at higher energy. Now if these properties are indeed correct, when a nucleon is split open in a high-energy collision its constituents—the quarks—should come out, since they are of low mass and

*Pauli's principle explained why electrons do not all stay in one single lowest-energy orbit. If they did, we would not have the periodic table of the elements, nor any (normal) stable matter.

have small interaction strength. Yet we have never seen free quarks. That is the paradox of the unseen quarks.

We now come to the experiments that established these strange properties of quarks. Consider two high-energy electron-positron ($e^- e^+$) collisions, one producing a muon pair ($\mu^- \mu^+$), the other a quark-antiquark pair ($q\bar{q}$). These collisions are known to go through two steps:

$$e^- + e^+ \rightarrow \text{virtual photon,} \tag{6}$$

$$\text{virtual photon} \rightarrow \begin{cases} \mu^- + \mu^+ \\ q + \bar{q} \rightarrow \text{hadrons.} \end{cases} \tag{7}$$

In the first step, the electron and positron annihilate into a "virtual" photon (γ) which, in the second step, converts into either $\mu^- + \mu^+$ or $q + \bar{q}$; the quark-antiquark pair, in turn, converts into all kinds of hadrons (pions, kaons, nucleons, and antinucleons). The amplitude of the first step is proportional to the electric charge of the electron, and of the second step to that of either the electric charge of the muon or that of the quark. Since the reaction rate is the square of the amplitude, we have for the ratio R of the two reaction rates in Equation (7):

$$R \equiv \frac{\Sigma \text{ rate } (e^- e^+ \rightarrow q\bar{q} \rightarrow \text{hadrons})}{\text{rate } (e^- e^+ \rightarrow \mu^- \mu^+)} = \Sigma\, Q_q^{\ 2} \tag{8}$$

where the summation sign Σ extends over all quark pairs that can be produced, and Q_q denotes the electric charge of the relevant quark (in units of the electric charge of the muon which, incidentally, is of the same magnitude as that of the proton).

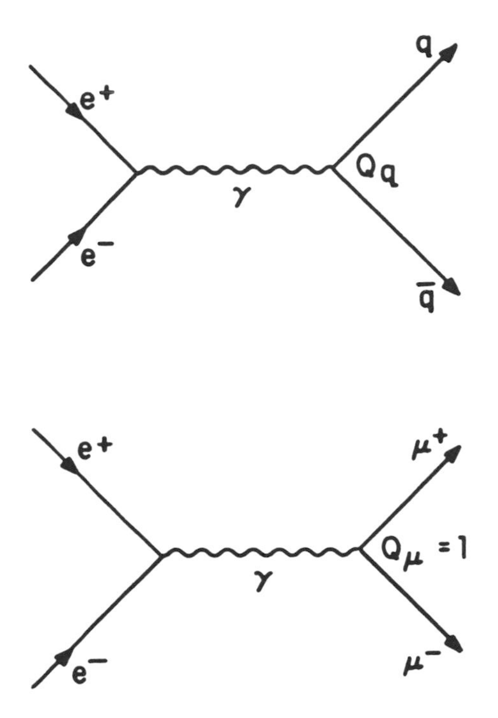

Figure 18. An electron-positron collision can produce either a quark-antiquark pair or a $\mu^-\mu^+$ ***pair.***

In Figure 18 we give a graphic representation of the reactions in Equations (6) and (7). The experimental result of the ratio R is plotted schematically versus the total (center of mass) energy E_{CM} in the e^- e^+ collision in Figure 19. We see that R is essentially flat, except for sudden jumps at certain critical energies. When E_{CM} is below 4 GeV, R is about 2; then it jumps to 3⅓ for E_{CM} between 4 and 10 GeV, and then it increases to 3⅔. This can be understood as follows.

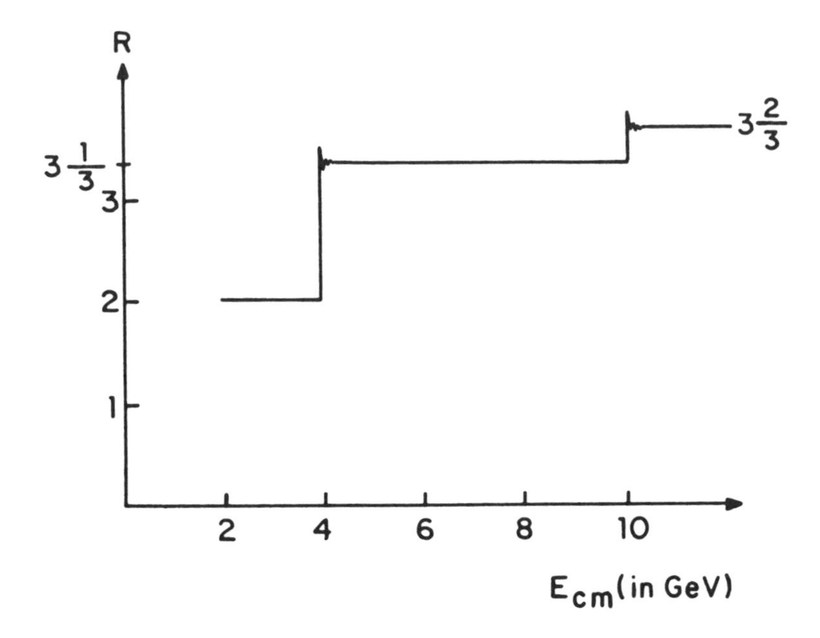

Figure 19. The ordinate R is the ratio of the production rate of hadrons vs. that of muon pairs in an electron-positron collision. The abscissa is the center-of-mass energy in units of 10^9eV.

For E_{CM} less than 4 GeV, a glance* at Table 2 tells us that only a $u\bar{u}$ pair, a $d\bar{d}$ pair, and an $s\bar{s}$ pair can be created; their corresponding electric charges Q_q are $2/3$, $-1/3$, and $-1/3$. Each reaction gives a term $Q_q{}^2$ to the sum in Equation (8). This gives

$$3 \times [(\tfrac{2}{3})^2 + (-\tfrac{1}{3})^2 + (-\tfrac{1}{3})^2] = 3 \times \tfrac{6}{9} = 2 \qquad (9)$$

*In order to produce a $q\bar{q}$ pair, the total available energy (the center-of-mass energy E_{CM}) must be larger than the combined masses of the quark and antiquark. From Table 2 we see that the combined masses of $u\bar{u}$, $d\bar{d}$, and $s\bar{s}$ are about 20 MeV, 10 MeV, and 200 MeV. Therefore, these quark-antiquark pairs can be produced when E_{CM} is less than 4 GeV (but larger than 200 MeV). To produce $c\bar{c}$ requires E_{CM} above 4 GeV.

where the first factor 3 is due to the three different species (colors) of each of the $u\bar{u}$, $d\bar{d}$, and $s\bar{s}$ pairs. When E_{CM} is above two times the mass of the c quark, the $c\bar{c}$ pair can be produced. This happens when E_{CM} is greater than 4 GeV. The opening of this new reaction adds to the overall R value by

$$3 \times (\tfrac{2}{3})^2 = 1\tfrac{1}{3}, \qquad (10)$$

making R increase from 2 to $3\frac{1}{3}$. In the above formula, the first factor 3 is again due to the three species, and the second factor $(2/3)^2$ is the square of the electric charge of the c quark. When E_{CM} is larger than 10 GeV, R suffers a further jump due to the opening of another new reaction: the creation of the $b\bar{b}$ quark; this in turn determines the mass of the b quark to be 5 GeV (half of 10 GeV). The increase this time is only $1/3$, which can be accounted for by multiplying the square of the b quark charge by the specie factor, 3:

$$3 \times (\tfrac{1}{3})^2 = \tfrac{1}{3}. \qquad (11)$$

These experiments determine the masses of the quarks, the magnitudes of their electric charges, and the existence of three color species. In Equation (8) we compare the quark pair production rate with the muon production rate. This makes sense only if the quark pairs and the muon pairs have similar types of interactions. Now the muons are known to have only weak interactions besides their electromagnetic interactions. But the quarks are supposed to have strong interactions, at least at the binding energy of the proton (about 1 GeV). That the R value experiment makes sense shows that the interaction between quarks must have become weak at higher energy. This property is referred to as "asymptotic freedom." If the quark mass is not heavy, and its apparent interaction strength is weak at high energy, how is it that in a high-energy collision we cannot see single quarks? Why can we not free the asymptotically free quarks?

A full understanding of this remarkable property of quarks requires the whole machinery of quantum chromodynamics (*QCD*). Again, the vacuum has to be involved. The mechanism to explain the unseen quarks is called "quark confinement," illustrated in Figure 20.

The vacuum in quantum chromodynamics is viewed as a complicated medium. The nucleons and mesons are visualized as bubbles (of radius about 1 fermi, 10^{-13} cm) in this medium. Inside each bubble, called a "bag," we have a quark-antiquark pair for the meson and three quarks for the nucleon. These bubbles are created out of the vacuum by the quarks. Because of the pressure exerted by the vacuum on the bubbles, quarks cannot come out, and this gives rise to the "quark confinement" phenomenon. Otherwise, the quarks move fairly freely within the bubble. The kinetic energy of the quarks withstands the pressure of the outside vacuum, preventing the bubble from collapsing.

Is it possible to test these ideas more directly? This leads again to the relativistic heavy-ion experiments discussed earlier. If one can collide, say, uranium on uranium at 100 GeV per nucleon, then we brew a high-energy mixture containing 1428 valence quarks and a sea of assorted quark-antiquark pairs and gluons. According to quantum chromodynamics, this will then create a much larger bubble of about ten-fermi diameter across. Inside this big bubble, each quark would behave even more like a free particle. In that sense one can excite the nuclear matter into a plasma of quarks and gluons, and thereby also change the background vacuum and further test the confinement prediction of *QCD*.

This then leads us back to the final theme of the preceding

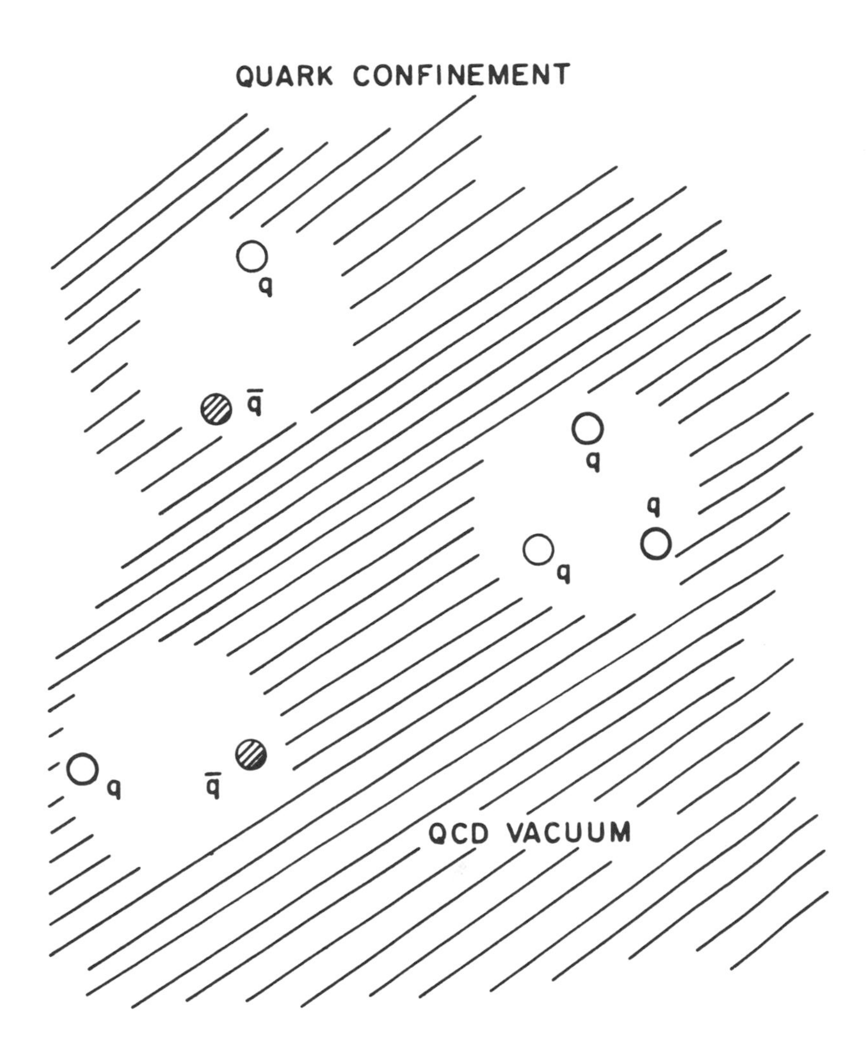

Figure 20. Nucleons (three-quark compounds) and mesons (quark-antiquark compounds) embedded in the vacuum according to quantum chromodynamics (QCD).

chapter. To resolve these two major puzzles, missing symmetries and unseen quarks, we invoke the dynamics of the vacuum. If the vacuum is indeed the underlying cause for these strange phenomena in the microscopic world of particle physics, it must also have been actively responsive to the macroscopic distribution of matter and energy in the universe. Because the vacuum is everywhere and forever, these two, the micro- and the macro-, have to be linked together; neither can be considered a separate entity.

In thinking on these matters, we might reason that if everything is composed only of particles, then our world should be a world of particles only. But most of us, living in the macroscopic environment, are quite unaware of the microscopic world. Nevertheless, the microscopic is the basic world, and the macroscopic only its manifestation. We are able to ignore this fact because our perceptions are not the most sensitive. Conversely, if these two do refer to the same world, how can we physicists often think of the realm of the microscopic as just a small isolated world consisting of a few small elementary particles? Yet this outlook permeates all our physics experiments and almost all our analyses. Then this, too, must be only an approximation; the perceived isolation cannot be real. To meditate on the union of the microscopic and the macroscopic is philosophy, to quantify their dualism is physics.

Appendix

The Four Groups of Symmetries

There are four main groups of symmetries, or broken symmetries, that are found to be of importance in physics:

1. Permutation symmetry: Bose-Einstein and Fermi-Dirac statistics.
2. Continuous space-time transformations, such as: translations, rotations, and accelerations.
3. Discrete transformations, such as: space inversion *P*, time reversal *T*, particle-antiparticle conjugation *C*, and *G*-parity.
4. Gauge transformations, which include: *U* (1) symmetries—conservation laws of electric charge, hypercharge, baryon number, and lepton number; *SU* (2) (isospin) symmetry; and *SU* (3) (color and flavor) symmetry.

The first group deals with the symmetry of identical particles (in the sense that all electrons are identical). Hence, the physical world must remain the same if we interchange (permute) any two electrons. This symmetry is therefore called permutation symmetry. Its consequences have been explored by Bose and Einstein for light quanta, and by Fermi and Dirac for electrons. The analysis of Bose and Einstein is also applicable to π mesons, K mesons, and gravitational quanta; that of Fermi and Dirac can also be applied to protons, neutrons, neutrinos, and μ mesons.

The second and third groups have already been explained in the text.

To describe the fourth group, gauge symmetry, one has

first to accept that the physical state of any collection of particles is described by a set of complex numbers ψ (the set is called a wave function), each of which carries both magnitude and phase. The magnitude of ψ can always be observed, because its square is the intensity of the particles. The observability of the phase of ψ underlies the whole basis of quantum mechanics. Gauge symmetry (associated with electricity) states, however, that the phase difference between two states of different charge can never be observed. Mathematically, this means that the physical world remains unchanged if we arbitrarily multiply ψ by a phase factor:

$$\psi \to \psi\, e^{iQ\theta},$$

where Q is the electric charge and θ is a real number. The consequence of this symmetry is the conservation of electric charge Q. The same idea can be applied to other relative phases, and that leads to the conservation of hypercharge, baryon number, and lepton number. Because $e^{iQ\theta}$ is a 1×1 unitary matrix, this symmetry is called $U(1)$ symmetry; sometimes it is also referred to as abelian symmetry, named after the Norwegian mathematician, N. H. Abel (1802–29). Generalization to 2×2 unitary matrices, or 3×3 unitary matrices, leads to $SU(2)$ or $SU(3)$ symmetry, sometimes referred to as nonabelian gauge symmetry. The $SU(2)$ symmetry, when applied to protons and neutrons, gives isospin conservation. The $SU(3)$ symmetry, when applied to different colors of quarks, forms the basis of quantum chromodynamics.

Among these four groups, the symmetries connected with the first two groups of transformations are, at present, believed to be exact. In the third group only the product *CPT* is perhaps exact, but each individual discrete symmetry operation is not. In the fourth group only some of the $U(1)$

Table 3. Examples of Symmetries in Physics

Nonobservables	Mathematical Transformations	Conservation Laws or Selection Rules
absolute spatial position	space translation $\vec{r} \rightarrow \vec{r} + \vec{\Delta}$	momentum
absolute time	time translation $t \rightarrow t + \tau$	energy
absolute spatial direction	rotation $\vec{r} \rightarrow \vec{r}\,'$	angular momentum
absolute right (or absolute left)	$\vec{r} \rightarrow -\vec{r}$	parity
absolute sign of electric charge	$e \rightarrow -e$	charge conjugation
absolute sign of time	$t \rightarrow -t$	time reversal
difference between identical particles	permutation	Bose-Einstein or Fermi-Dirac statistics
relative phase between states of different electric charge Q	$\psi \rightarrow e^{iQ\theta}\psi$ (gauge transformation)	electric charge

and the color $SU(3)$ symmetries are thought to be exact; this group is also referred to as unitary symmetries, because of their close association with the unitary matrices in mathematics.

The same theme—nonobservables, invariance under certain mathematical transformations, and conservation laws—goes through each of these symmetry principles. These are illustrated in Table 3.

References

(A far-from-complete selection,
arranged in the order of topics discussed.)

General Reference Books

J. Bernstein, *A Comprehensible World: On Modern Science and Its Origins,* New York, Random House, 1967.

G. Feinberg, *Solid Clues,* New York, Simon and Schuster, 1985.

Y. Nambu, *Quarks,* Singapore, World Scientific Publishing Company, 1985.

H. Pagels, *The Cosmic Code,* New York, Simon and Schuster, 1982.

A. Pais, *Inward Bound,* New York, Oxford University Press, 1986.

A. Zee, *Fearful Symmetry,* New York, Macmillan Publishing Company, 1986.

Parity Nonconservation

T. D. Lee and C. N. Yang, "Questions of Parity Conservation in Weak Interactions," Phys. Rev. *104,* 254 (1956).

C. S. Wu, E. Ambler, R. W. Hayward, D. D. Hoppes, and R. P. Hudson, "Experimental Test of Parity Conservation in Beta Decay," Phys. Rev. *105,* 1413 (1957).

R. L. Garwin, L. M. Lederman, and M. Weinrich, "Observation of the Failure of Conservation of Parity and Charge Conjugation in Muon Decays: The Magnetic Moment of the Free Muon," Phys. Rev. *105,* 1415 (1957).

V. L. Telegdi and A. M. Friedman, "Nuclear Emulsion Evidence for Parity Nonconservation in the Decay Chain $\pi^+ - \mu^+ - e^+$," Phys. Rev. *105,* 1681 (1957).

CP and T Violations

T. D. Lee, R. Oehme, and C. N. Yang, "Remarks on Possible Noninvariance under Time Reversal and Charge Conjugation," Phys. Rev. *106,* 340 (1957).

J. H. Christenson, J. W. Cronin, V. L. Fitch, and R. Turlay, "Evidence for the 2π Decay of the $K_2{}^0$ Meson," Phys. Rev. Lett. *13,* 138 (1964).

S. Bennett et al., "Measurement of the Charge Asymmetry in the Decay $K_L{}^0 \rightarrow \pi^{\pm} + e^{\mp} + \nu$," Phys. Rev. Lett. *19,* 993 (1967).

Two-component Neutrino

T. D. Lee and C. N. Yang, "Parity Nonconservation and a Two-Component Theory of the Neutrino," Phys. Rev. *105,* 1671 (1957).

L. Landau, "On the Conservation Laws for Weak Interactions," Nucl. Phys. *3,* 127 (1957).

A. Salam, "On Parity Conservation and Neutrino Mass," Nuovo Cimento *5,* 299 (1957).

CPT Theorem

W. Pauli, "Exclusion Principle, Lorentz Group and Reflection of Space-time and Charge," in *Niels Bohr and the Development of Physics,* ed. W. Pauli, L. Rosenfeld, and V. Weisskopf, New York, McGraw-Hill, 1955.

Spontaneous Symmetry Breaking

Y. Nambu, "Axial Vector Current Conservation in Weak Interactions," Phys. Rev. Lett. *4,* 380 (1960).

J. Goldstone, "Field Theories with 'Superconductor' Solutions," Nuovo Cimento *19,* 154 (1961).

P. W. Higgs, "Broken Symmetry, Massless Particles and Gauge Fields," Phys. Lett. *12,* 132 (1964); "Broken Symmetries and the Masses of Gauge Bosons," Phys. Rev. Lett. *13,* 508 (1964); "Spontaneous Symmetry Breakdown without Massless Bosons," Phys. Rev. *145,* 1156 (1966).

Intermediate Bosons

T. D. Lee, M. Rosenbluth, and C. N. Yang, "Interaction of Mesons with Nucleons and Light Particles," Phys. Rev. *75,* 905 (1949).

S. Weinberg, "A Model of Leptons," Phys. Rev. Lett. *19,* 1264 (1967).

G. Arnison *et al.*, "Experimental Observation of Isolated Large Transverse Energy Electrons with Associated Missing Energy at $\sqrt{s}$= 540 GeV," Phys. Lett. *122*B, 103 (1983); "Experimental Observation of Lepton Pairs of Invariant Mass around 95 GeV/e at the CERN SPS Collider," Phys. Lett. *126*B, 398 (1983).

M. Banner *et al.*, "Observation of Single Isolated Electron of High Transverse Momentum in Events with Missing Transverse Energy at the CERN p$\bar{\text{p}}$ Collider," Phys. Lett. *122*B, 476 (1983).

P. Bagnaia *et al.*, "Evidence of $Z^0 \rightarrow e^+ e^-$ at the CERN p$\bar{\text{p}}$ Collider," Phys. Lett. *129*B, 130 (1983).

Neutral Current

F. J. Hasert *et al.*, "Search for Elastic Muon-Neutrino Electron Scattering," Phys. Lett. *46*B, 121 (1973).

The Standard Model

S. Weinberg, "A Model of Leptons," Phys. Rev. Lett. *19,* 1264 (1967).

A. Salem, "Weak and Electromagnetic Interactions," *Proceedings of the Eighth Nobel Symposium,* ed. N. Svartholm, New York, Wiley-Interscience, 1968, p. 367.

S. L. Glashow, "Partial-Symmetries of Weak Interactions," Nucl. Phys. *22,* 579 (1961).

A. Salam and J. C. Ward, "Weak and Electromagnetic Interactions," Nuovo Cimento *11,* 568 (1959); "Electromagnetic and Weak Interactions," Phys. Lett. *13,* 168 (1964).

The Michel Parameter

L. Michel, "Interaction between Four Half-Spin Particles and the Decay of the Mu Meson," Proc. Phys. Soc. (London) A*63*. 514 (1950).

Quarks and Colors

G. Zweig, CERN report (unpublished).

M. Gell-Mann, "A Schematic Model of Baryons and Mesons," Phys. Lett. *8,* 214 (1964).

O.W. Greenberg, "Spin and Unitary-Spin Independence in a Paraquark Model of Baryons and Mesons," Phys. Rev. Lett. *13,* 598 (1964).

J/ψ and the c Quark

J. J. Aubert *et al.*, "Experimental Observation of a Heavy Particle J," Phys. Rev. Lett. *33,* 1404 (1974).

J. E. Augustin *et al.*, "Discovery of a Narrow Resonance in e e Annihilation," Phys. Rev. Lett. *33,* 1406 (1974).

Upsilon and the b Quark

S. W. Herb *et al.*, "Observation of a Dimuon Resonance at 9.5 GeV in 400 GeV Proton-Nucleus Collisions," Phys. Rev. Lett. *39,* 252 (1977).

ν_e and ν_μ

G. Danby, J.-M. Gaillard, K. Goulianos, L. M. Lederman, N. Mistry, M. Schwartz, and J. Steinberger, "Observation of High-Energy Neutrino Reaction and the Existence of Two Kinds of Neutrinos," Phys. Rev. Lett. *9,* 36 (1962).

τ and ν_τ

M. L. Perl *et al.*, "Evidence for Anomalous Lepton Production in e^+-e^- Annihilation," Phys. Rev. Lett. *35*. 1489 (1975).

Non-Abelian Gauge Theory

C. N. Yang and F. Mills, "Conservation of Isotopic Spin and Isotopic Gauge Invariance," Phys. Rev. *96,* 191 (1954).

O. Klein, "On the Theory of Charged Fields," in *New Theories in Physics* (International Institute of Intellectual Cooperation, League of Nations, 1938), p. 77.

Quantum Chromodynamics and Asymptotic Freedom

H. D. Politzer, "Reliable Perturbative Results for Strong Interactions," Phys. Rev. Lett. *30,* 1346 (1973).

D. Gross and F. Wilczek, "Ultraviolet Behavior of Non-Abelian Gauge Theories," Phys. Rev. Lett. *30,* 1343 (1973).

G. 't Hooft, talk at the Marseilles meeting, 1972 (unpublished).

Bag Model

A. Chodos, R. J. Jaffe, K. Jonson, C. B. Thorn, and V. F. Weisskopf, "New Extended Model of Hadrons," Phys. Rev. D*9,* 3471 (1974).

Electron-Positron Collision

See the *Proceedings of the 1985 International Symposium on Lepton and Photon Interactions at High Energies,* Kyoto University, Kyoto, 1986.